The Antarctic Ocean

by Anne Ylvisaker

Consultant:
Sarah E. Schoedinger
Education Coordinator
Consortium for Oceanographic Research and Education
Washington, D.C.

Bridgestone Books
an imprint of Capstone Press
Mankato, Minnesota

Bridgestone Books are published by Capstone Press
151 Good Counsel Drive, P.O. Box 669, Mankato, Minnesota 56002
http://www.capstone-press.com

Library of Congress Cataloging-in-Publication Data
Ylvisaker, Anne.
    The Antarctic Ocean / by Anne Ylvisaker.
    p. cm.—(Oceans)
    Includes bibliographical references and index.
    ISBN 0-7368-1420-5 (hardcover)
    1. Oceanography—Antarctic Ocean—Juvenile literature. 2. Antarctic Ocean—
Juvenile literature. I. Title.
GC461 .Y58 2003
551.46′9—dc21   2002000691

Summary: Introduces the southernmost ocean and provides an activity for measuring
    the coldness of ice.

Editorial Credits
Megan Schoeneberger, editor; Karen Risch, product planning editor; Linda Clavel,
    designer; Image Select International, photo researcher

Photo Credits
Art Directors and TRIP/E. Smith, 14; Corbis Images/PictureQuest, cover; Corbis/Rick
Price, 20; Digital Wisdom/Mountain High, 6, 8 (map); Doug Allan/Science Photo
Library, 18; Erin Scott/SARIN Creative, 10; ImageState, 4, 12, 16; RubberBall
Productions, 22, 23; Science Photo Library, 8 (photo)

1 2 3 4 5 6 07 06 05 04 03 02

Table of Contents

The Antarctic Ocean

The Antarctic Ocean is the fourth largest ocean. It is larger than North America. The Antarctic Ocean covers about 8 million square miles (21 million square kilometers). It also is the most southern of all oceans. Some scientists call it the Southern Ocean.

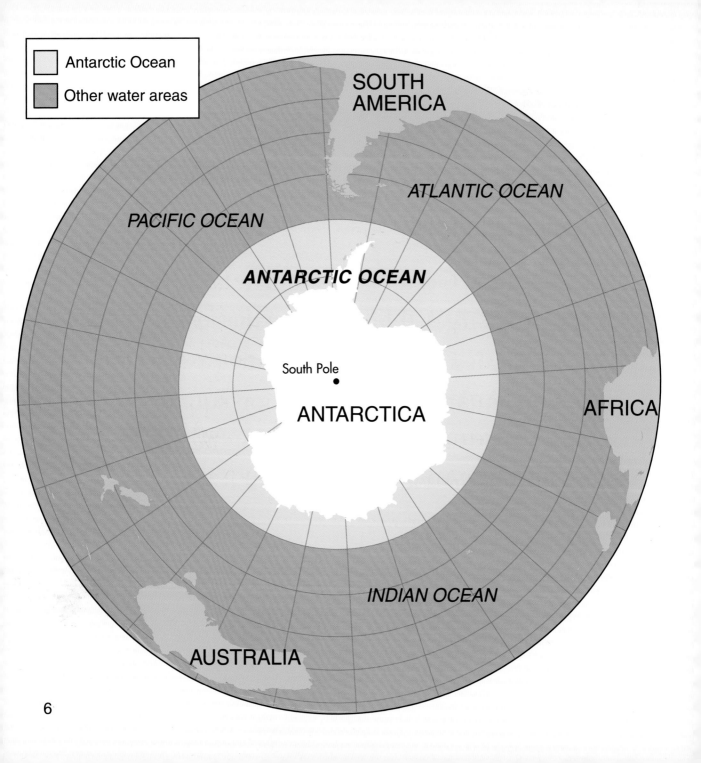

The Location of the Antarctic Ocean

The Antarctic Ocean circles the continent of Antarctica. Antarctica is at the southern end of Earth. The South Pole is in Antarctica. The Indian Ocean, Atlantic Ocean, and Pacific Ocean all flow into the Antarctic Ocean.

continent
one of the seven main landmasses of Earth

South
Sandwich
Trench

ANTARCTIC OCEAN

South Pole
•

Ocean Depths

deepest   shallowest

An icebreaker is a ship that
can break through ice.
Scientists use icebreakers to
study parts of the Antarctic
Ocean. Some icebreakers
can break through ice that is
23 feet (7 meters) thick.

The Depth of the Antarctic Ocean

The average depth of the Antarctic Ocean is 10,627 feet (3,239 meters). This depth is about 2 miles (3 kilometers). The deepest place in the Antarctic Ocean lies near South Sandwich Trench. There, the ocean is more than 5 miles (8 kilometers) deep.

depth
a measure of how deep something is

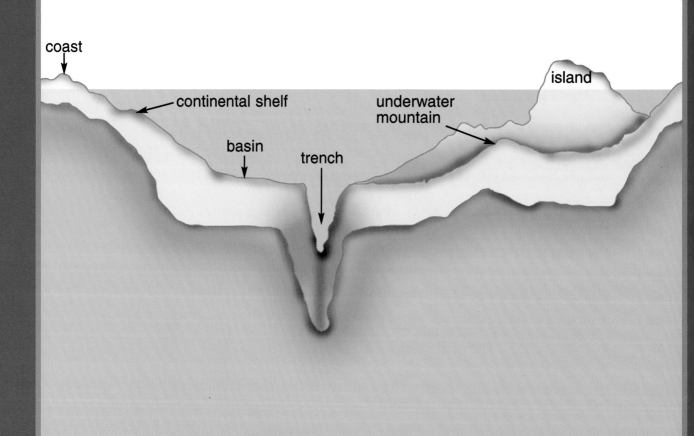

coast

island

continental shelf

underwater mountain

basin

trench

10

The Bottom of the Antarctic Ocean

The continental shelf is the shallowest part of an ocean's floor. It slopes from the coast to the basin. The basin is mostly flat. It has some mountains, volcanoes, hills, and trenches. Thick mud covers the Antarctic Ocean floor.

basin
the low, flat part
of an ocean's floor

Icebergs are large pieces of floating ice. They break off from slow-moving pieces of frozen freshwater called glaciers. The largest iceberg seen in the Antarctic Ocean was about 200 miles (320 kilometers) long.

The Water in the Antarctic Ocean

The water in the Antarctic Ocean is cold and salty. The average surface temperature of the Antarctic Ocean is 32 degrees Fahrenheit (0 degrees Celsius). Ice and icebergs cover the Antarctic Ocean most of the year.

surface
the top or outside layer of something

Fun Fact
Summer in Antarctica happens from November to January. Winter happens from May to September. Antarctica is dark for many months during winter.

The Climate around the Antarctic Ocean

Antarctica is cold and windy. Average winter temperatures are 22 degrees Fahrenheit (30 degrees Celsius) below zero. Winds can blow more than 190 miles (306 kilometers) per hour.

climate
the usual weather that occurs in a place

Fun Fact
Penguins cannot fly. These birds use their wings like flippers to swim in the ocean. They spend most of their time in water.

16

Animals in the Antarctic Ocean

Many animals live in the Antarctic Ocean.
Whales swim in the Antarctic Ocean.
Penguins and seals live at the edge of the
ocean. Krill live in the Antarctic Ocean.
These tiny shrimplike animals are food
for many other animals. Petrels and other
birds fly over the water looking for fish.

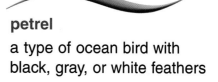

petrel
a type of ocean bird with
black, gray, or white feathers

This photo of phytoplankton was taken through a microscope. A microscope makes very small objects appear large enough to study.

Plants in the Antarctic Ocean

Most plants do not grow well in cold and dark areas. Phytoplankton are the only plants in the Antarctic Ocean. These tiny plants are food for many ocean animals. Phytoplankton in the Antarctic Ocean grow only during summer.

phytoplankton
tiny plants that drift
in oceans

20

Keeping the Antarctic Ocean Healthy

People are working to protect the Antarctic Ocean and its animals. Whales and seals nearly became extinct from hunting. Now, hunting most whales and seals is against the law.

extinct
no longer living
anywhere in the world

Hands On: How Cold Is Cold?

Floating ice makes Antarctic water cold near the surface.
You can try this experiment to see how cold the water can be.

What You Need

Two bowls
Water from a faucet
Kitchen thermometer
Ice cubes

What You Do

1. Fill one bowl with very cold water.
2. Check the water's temperature with the thermometer. It should be 32 degrees Fahrenheit (0 degrees Celsius). If necessary, add ice cubes until the water is cold enough. This temperature is how cold the water in the Antarctic Ocean can be.
3. Fill the other bowl with warm water.
4. Check the temperature with the thermometer. Try to fill the bowl with water that is around 75 degrees Fahrenheit (24 degrees Celsius). This temperature is the temperature of the water in the Atlantic Ocean where people swim.

Put one hand in each bowl to compare the temperature of the two oceans.

Words to Know

average (AV-uh-rij)—the most common amount of something; an average amount is found by adding figures together and dividing by the number of figures.

continental shelf (KON-tuh-nuhn-tuhl SHELF)—the shallow area of an ocean's floor near a coast

krill (KRILL)—very small animals that look like shrimp; krill are the main food source for many ocean animals.

phytoplankton (FITE-oh-plangk-tuhn)—tiny plants that drift in oceans; phytoplankton are too small to be seen without a microscope.

protect (pruh-TECT)—to keep safe

shallow (SHAL-oh)—not deep

trench (TRENCH)—a long, narrow valley in an ocean

volcano (vol-KAY-noh)—a mountain with vents; melted rock flows out of the vents when a volcano erupts.

Read More

Bagley, Katie. *Antarctica.* Continents. Mankato, Minn.: Bridgestone Books, 2003.

Fowler, Allan. *Antarctica.* Rookie Read-About Geography. New York: Children's Press, 2001.

Penny, Malcolm. *The Polar Seas.* Seas and Oceans. Austin, Texas: Raintree Steck-Vaughn, 1997.

Stone, Lynn M. *The Antarctic Ocean.* Antarctica. Vero Beach, Fla.: Rourke, 1995.

Internet Sites

Oceanlink
http://www.oceanlink.island.net
What's It Like Where You Live?—Temperate Oceans
http://mbgnet.mobot.org/salt/oceans

Index